Student Workbook for:

The Reason

What Science Says About God

By Robert J. Spitzer, S.J., Ph.D.
and Claude LeBlanc, M.A.

MAGIS CENTER
OF REASON AND FAITH
www.magisreasonfaith.org

Cover art by Jim Breen

Photos:

Photos from the Hubble Telescope: courtesy NASA, PD-US
Stephen Hawking (page 13): courtesy NASA, PD-US
Galileo (page 17): PD-Art
Arno Penzias (page 21): courtesy Kartik J, GNU Free Documentation License
Aristotle (page 31): courtesy Eric Gaba, Creative Commons
Albert Einstein (page 37): PD-Art
Georges Lemaitre (page 37): PD-Art
Alexander Vilenkin (page 39): courtesy Lumidek, Creative Commons
Alan Guth (page 39): courtesy Betsy Devine, Creative Commons
Roger Penrose (page 49): PD-Art
Paul Davies (page 49): courtesy Arizona State University, PD-Art
Sir Arthur Eddington (page 63): courtesy Library of Congress, PD-US

This edition published by:

Magis Publications
2532 Dupont Drive
Irvine, California 92612

www.magisreasonfaith.org

ISBN: 978-0-9838945-1-3

Printed in the United States of America

"It is said that an argument is what convinces reasonable men and a proof is what it takes to convince even an unreasonable man. With the proof now in place, cosmologists can no longer hide behind the possibility of a past-eternal universe... There is no escape, they have to face the problem of a cosmic beginning."

From Many Worlds in One: The Search for Other Universes
(New York Macmillan) 2007, by Dr. Alexander Vilenkin,
professor of theoretical physics and director of the Institute of Cosmology
at Tufts University

Table of Contents

Introduction

Welcome to ***The Reason: What Science Says about God.*** This video course was created to give answers to some of the toughest questions you will ever face in your life.

- In a world full of scientific discoveries, is God still relevant?
- Should you believe in both God and science?
- What are the limits of science?
- How should you respond if someone challenges your faith using science?
- Is there actually any evidence beyond the Bible that God is real?

Most of you taking this course come from a background where you have been taught that God is real and present in your lives. You may have questioned your faith, and that's not a bad thing. Mature faith grows from such questioning, and you shouldn't be afraid to ask tough questions. Much of what you've been taught, however, has probably been about how to be a good Christian, how to worship God, or how to be a part of the Church community. All of this training starts with the basic assumption that God exists. Not everyone believes God is real.

So what do you do if you aren't so sure God actually exists? Sooner or later, you will run into intelligent people who will argue that God is not real. Many of them will use science as the basis for their assertions. Some of them will try to convince you, too.

Perhaps they already have.

This series explores what science really says about God. It is based on the latest discoveries in astrophysics and philosophy, and the advisors behind this course are Nobel and Templeton laureates. They work for NASA. They're eminent university professors. They are among the greatest scientific minds of our generation.

They also believe in God. Why?

Because that's what the evidence tells them. In fact, many physicists who were once atheists now believe in God because a creator is the most scientifically probable explanation for our universe. You may find their conclusions startling at first, but you will also find their logic and evidence to be quite comprehensive. Believing in God makes a great deal more sense than not believing in God - mainstream science tells us this.

God is real. This course explores the evidence.

Notes

Segment 1
Can Science Disprove God?

Objectives

Students will learn:

1. The scientific method: how it works, what it can prove, and what its limits are;

2. Why science is incapable of disproving things that are beyond the physical universe, such as God; and

3. How science indicates our universe had a beginning, implying that something outside of it had to start it, giving evidence for a transcendent being beyond the universe.

Overview

We meet Joe (a college freshman) who is asking Alana (working on her doctorate in philosophy) and Dan (in a Ph.D. program in physics) for help in dealing with some "scientific" challenges that Tyler (Joe's atheist roommate) is raising about his belief in God. Together, they discuss how science works, what its limits are in proving or disproving things (including God) and how else (besides scripture or theology) we can obtain evidence for God. Using the fact that our universe had a beginning, and considering that it could not have caused itself, they are left with the conclusion that something outside of the universe and the laws of physics must have created it. They conclude that not only is science unable to disprove God's existence, but it may actually contribute to making belief in him more reasonable.

Scientific Summary of Segment 1

Can Science Disprove God?

1. **What is the scientific method?**

The scientific method is a tool scientists use to understand the natural world. This method is usually made-up of the following steps:

a. **Asking a question,**
b. **Doing background research,**
c. **Constructing a hypothesis,**
d. **Testing your hypothesis by doing an experiment,**
e. **Analyzing your data and drawing a conclusion, and**
f. **Communicating your results.**

After scientists share their results, other scientists will duplicate the experiment to verify the results. If an experiment can be verified, it lends credibility to the results. Even though scientists may reach different conclusions regarding the significance of the results, the results are considered reliable if repeatable.

2. **Is the scientific method useful for disproving God's existence?**

No, because the scientific method is limited to using empirical observation (within the universe), and its experiments must be measurable. However, God is not in the universe and is not observable, and, furthermore, he is not measurable (because he is unrestricted in power and not conditioned by either space or time). The two reasons God's existence cannot be disproven by the scientific method are:

a. **It is nearly impossible to completely disprove something using the scientific method.** While the scientific method can easily be used to prove something, to disprove something scientists would have to completely rule it out everywhere it might exist (at the same time). For example, to disprove the existence of alien life in our universe, scientists would have to completely rule out their presence in every place all at once (because the aliens might be hiding from the scientists, staying one planet ahead of them).

b. **God is transcendent (existing outside of the universe), and can't be disproved by a method limited to the universe, human observation, or measurement.**

For these two reasons atheists cannot use scientific evidence to disprove God's existence in support of their atheism. As Richard Dawkins recently admitted, "...science cannot disprove God."

3. **Can the scientific method be useful for proving God's existence?**

Yes, in two ways:

a. **If science can show that the universe had a beginning in time, then it would imply a creator.** Why? Consider the following—if the universe had a beginning (a point at which it came into existence) then prior to that beginning it would have been nothing. So what can nothing do? As you may have surmised, nothing can only do nothing. Now—if nothing can only do nothing, then when the universe was nothing, it could not have made itself something (because it could only do nothing). This means that something beyond the universe—like God—would have to have brought it into existence (moved it from nothing to something).

b. **If science can show that the conditions and constants of the universe being able to support life are highly improbable (and that a multiverse explanation is not really adequate), it is reasonable to conclude that a superior intelligence—like God—designed it to be that way.**

These points will be further explained in **Segment 2** and **Segment 3**.

Notes

Handout 1a—Video Review and Discussion

Can Science Disprove God?

Opening Prayer

Oh Lord, we ask you to help us appreciate science as a way to know about the universe you created. Give us an understanding of how science works and what it is able to do, so that we may know what it has learned. Give us the clarity to know what is real from what is just an opinion, so that we may come to know the truth. Amen.

Opening Reflection and Sharing

What do you think: *Can science disprove God?*

Review Questions

1. What evidence does Tyler use to make his claim that, "Science has disproven God?"

2. What kind of evidence is Joe looking for to show Tyler he is mistaken?

3. What is the scientific method, and what can it claim to do?

4. Using the example of aliens, what are the limits of science in being able to prove or disprove things?

Name: **Date:** **Period:**

5. How do the limits of science apply to disproving God's existence?

6. What is the current scientific evidence that shows the existence of God?

7. Other than the Big Bang Theory, what other evidence is there from science that makes the existence of God more reasonable?

8. What is the video's response to Tyler's claim that, "Science has disproven God?"

Class Discussion Questions

1. Inasmuch as science is limited in its search for answers to measurable things in the physical universe, what are some things (other than God) that science is unable to prove or disprove?

2. Why do you think some scientists claim that God is incompatible with science, even though this is apparently not the case?

3. Do you think that science provides, or may be able to provide, evidence that makes belief in God more reasonable? Why or why not?

4. What previous beliefs about science and/or faith have changed for you because of this video? Why have they changed?

Closing Prayer

Thank you, Lord, for showing us what science is, how it works, and how important it is in understanding the universe you created. We appreciate that science can even help us to know that you exist and that the way the universe works reveals that you are more intelligent than we could ever imagine. May we come to more fully appreciate the way you work in the world and in our lives each day. Amen.

Handout 1b

Hawking is Talking Again, But What is He Saying?

Directions: Read the following quotes from Professor Stephen Hawking and complete the assignment below.

In 1988, in *A Brief History of Time*, Hawking's most famous work, he seems to recognize the limits of science and the necessity for a creator. In it he wrote: "If we discover a complete theory, it would be the ultimate triumph of human reason – for then we should know the mind of God." And he went on to say that: "Even if there is only one possible unified theory, it is just a set of rules and equations. What is it that breathes fire into the equations and makes a universe for them to describe?"

However, in his latest book, *The Grand Design*, Hawking said something remarkably different: "Because there is a law such as gravity, the universe can and will create itself from nothing. Spontaneous creation is the reason there is something rather than nothing, why the Universe exists, why we exist." He added: "It is not necessary to invoke God to light the blue touch paper and set the universe going."

Assignment: Based on the scientific and philosophical evidence presented in ***The Reason*, Segment 1**, and the information you collected from the summary, complete this short reflection on Hawking's change of opinion about the universe's need for a creator:

> *Although Dr. Hawking is a well-respected scientist, and deservedly so, his philosophical views are in question. He makes a basic error in thinking that leads him to reach a false conclusion.*

Name: **Date:** **Period:**

Notes

Handout 1c

What Can Galileo's Story Teach Us About Science and God?

Directions: Download and listen to the podcast by Prof. Thomas E. Woods, Jr. titled "The Galileo Files" http://www.ewtn.com/vondemand/audio/seriessearch prog.asp?seriesID=7129&T1=

(This link is courtesy of EWTN Audio Library, approximately 25 minutes long.)

Content Review

1. Was the charge against Galileo that he subscribed to a scientific theory that was different from the Bible?

2. Was the Church against the heliocentric theory? Explain.

3. Did the opponents of Galileo have scientific and mathematical arguments which were valid according to the astronomical instrumentation of the day?

Name: **Date:** **Period:**

4. Was Galileo supported by the Church prior to the publication of his *Dialogue Concerning the Two Chief World Systems*?

5. Was the Church ever anti-science prior to, during, or after Galileo's trial?

Notes

Name: Date: Period:

The Reason: Segment 1 Quiz

Can Science Disprove God?

Modified True or False

If the answer is true, mark true, but if it is false, mark false and re-write the sentence to be a true statement.

Example:

_____ A. ***The Reason*** is a video series that attempts to demonstrate that science and faith are incompatible.

Answer:

False A. ***The Reason*** is a video series that attempts to demonstrate that science and faith are ~~incompatible~~. compatible.

_____ 1. Scientific evidence now exists that proves the universe was either not created or that the universe created itself, making God unnecessary.

_____ 2. The scientific method is a tool used by scientists to find answers about things that are testable within our physical universe.

_____ 3. It is certain that science will someday absolutely prove whether or not aliens exist in our universe.

_____ 4. It is much easier for science to "prove" something than to "disprove" something.

_____ 5. Science may one day be able to disprove God's existence.

_____ 6. The theory of evolution has given us the best evidence from science to believe that the existence of God is reasonable.

_____ 7. Many philosophers have been able to demonstrate that there must be an infinite creator of some kind behind our universe.

_____ 8. Philosophy tells us that something like God would have to always exist in order for anything else to exist.

_____ 9. Physicists have found evidence that the universe is fine-tuned (against all reasonable odds) for life to exist, meaning that it is much more likely that it happened by chance instead of being created.

_____ 10. This segment of ***The Reason*** concludes that there is no reason to become an atheist because of scientific evidence.

Meet the Scientists

Arno Penzias

Dr. Arno Penzias was born in 1933 in Germany. He fled with his family at the age of six to the United States to escape the Nazis. He became a U.S. citizen in 1946 and earned his Ph.D. in physics in 1962 from Columbia University. In 1964, with Robert Wilson, Penzias encountered unexplained radio noise coming equally from every part of the sky while using the Bell Labs radio telescope in Holmden, New Jersey. They realized it was Cosmic Microwave Background Radiation remaining from the Big Bang, confirming that it had occurred. Both Penzias and Wilson received the 1978 Nobel Prize in Physics.

Notes

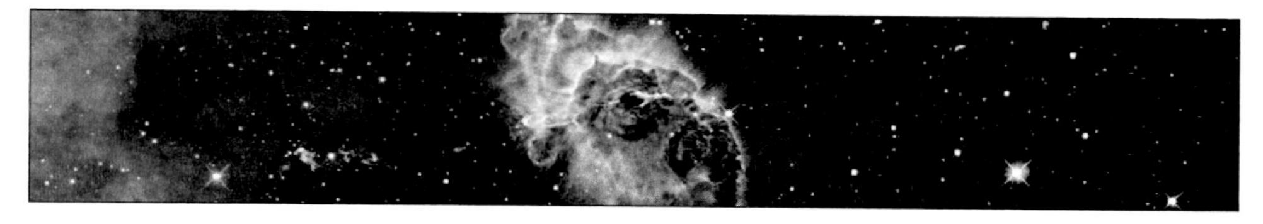

Segment 2
Is There Any Evidence for a Creator in the Universe?

Objectives

Students will learn:

1. That philosophy offers logical proofs for an eternally-existing being outside of the universe which cannot be disproved by science;

2. That there is strong and persuasive theory and evidence for the Big Bang, and that all competing theories about the universe are speculative;

3. The significance of the BVG Theorem: Any possible universe which is expanding must have a beginning, and, therefore, must be created; and

4. Other evidence for a beginning and creation of our universe.

Overview

Joe visits Alana and Dan again to report on how his conversations with his roommate, Tyler, are going. While Tyler now admits that science cannot disprove God's existence, he is adamant that there can never be any so-called proofs for God from philosophy that science can't disprove and that science could never offer any evidence that would support belief in God either. Alana explains that since everything in the universe (and the universe itself) had a beginning, it means that something outside of the universe must have created them. That creator would have had to have always existed. Dan shows Joe the theory and evidence for the expansion of the universe, meaning that it most probably had a beginning. They even demonstrate that current challenges to the Big Bang Theory (such as the bouncing universe or the multiverse) would need a beginning, too, according to the BVG Theorem, since they are all expanding universes. So, the Big Bang Theory—or any of its alternatives—requires a creator.

Scientific Summary of Segment 2

Is There Any Evidence for a Creator in the Universe?

1. **What is the Big Bang Theory?**

   The Big Bang Theory is considered one of the most reliable scientific theories by the vast majority of physicists. It holds that the universe began as a very small point and expanded like an inflating balloon, causing its matter to expand as well. Once galaxies formed, the universe's expansion causes them to continue to move away from each other.

2. **Who developed the Big Bang Theory, and what was the reaction of Albert Einstein?**

   It was a Belgian priest, Fr. Georges Lemaitre—also a physicist, having received a Ph.D. from MIT—who proposed the theory in response to an unanswered question in Einstein's General Theory of Relativity. Einstein eventually accepted Lemaitre's theory, admitting it was one of the most satisfying explanations of creation he had ever heard, but only after it was verified by observational evidence gathered by Edwin Hubble and others.

3. **What have we learned about the origin of universe from science?**

   Using evidence from many different sources, science has verified that the universe began with a Big Bang about 13.7 billion years ago.

   a. **Hubble's 1929 discovery of red shifts (evidence that galaxies are expanding away from one another),**
   b. **Penzias' and Wilson's 1963 discovery of a microwave radiation evenly distributed throughout the universe (which is expected from the Big Bang), and**
   c. **Evidence from the COBE Satellite (launched in 1989) and WMAP Satellite (launched in 2001, measuring the early universe).**

4. **What else has science told us about the universe?**

   a. **The rate at which the universe is expanding is increasing over time, and**
   b. **There is a finite (limited) amount of matter in the universe made up of:**

      1. Visible matter (4.6%),
      2. Dark matter (23%), and
      3. Dark energy (72.4%).

      - Visible matter is capable of electromagnetic and luminescent activity. There are about 10^{80} baryons (protons and neutrons) weighing approximately 10^{55} kg in our universe.

- Dark matter performs no known electromagnetic or luminescent activities; however, it strongly interacts with gravity and holds galaxies together while intergalactic space continues to expand (stretch) at an ever increasing rate.
- Dark energy is not like dark matter. In fact, it has the opposite effect. Dark energy is like a field which attaches itself to the space time field and causes the space time field to expand at ever greater rates.

5. Was the Big Bang the beginning of the universe?

Most physicists think it was, but others have suggested two alternatives:

a. **The bouncing-universe hypothesis:** Some physicists speculate that the universe might be expanding from a previous cycle which collapsed after having expanded for a period of time. Conceivably, there could be an infinite number of these cycles extending for an infinite amount of time in the past.

b. **The multiverse hypothesis:** Other physicists speculate that the universe might be just one bubble universe amidst trillions upon trillions of other bubble universes in a mega-universe called a "multiverse." This could have conceivably existed for an infinite amount of time.

However, there is no evidence for either of these hypotheses. These alternatives to the Big Bang Theory are often suggested in order to eliminate the need for an intelligent creator; but, as we will see, they do not.

6. Would a bouncing-universe, or a multiverse, need a beginning?

Yes, using space-time geometry proofs (which can be formulated from the general physics of space-time fields), physicists have given proofs that both of these speculative alternatives to the Big Bang Theory would themselves need a beginning. (In mathematics, a proof means that if certain conditions are met, other conditions are also true.) Two of these proofs, as well as a study of the various models of expanding universes used to make this claim, are:

a. **The 1993 Borde-Vilenkin Proof:** This proof (of physicists Dr. Arvin Borde and Dr. Alexander Vilenkin – see "meet the physicists in this section") shows that if any expanding universe meets five conditions, it would have a beginning. There is one possible exception to this proof, but it is highly highly improbable that it would apply to our universe or any universe connected to ours (including a bouncing-universe or a multiverse). MIT physicist Dr. Alan Guth, the "father of inflationary theory," does not think this is an important exception.

b. **The 1999 Guth study of inflationary model universes:** Guth's study shows that even though physicists have worked very hard to construct a model of an expanding universe that doesn't have a beginning, they have been unsuccessful—none of these models can be eternal into the past, and they must all have a beginning.

c. **The 2003 BVG Theorem:** This is considered the most important proof (named after Borde, Vilenkin, and Guth) because it has only one condition requiring a beginning—that the average rate of expansion of a universe is greater than zero. This means that in a universe's lifetime, it expands more than it contracts.

Every universe which could be connected with ours meets this condition, including **a multiverse**, (because in order to create bubble-universes it must be inflating) and **a bouncing-universe** (because it must expand before it collapses). No exceptions to this proof have been found, and it is believed that one may never be found (because it is very difficult to find exceptions to proofs with only one condition).

7. How do these proofs show the necessity of a beginning for any universe?

Vilenkin explained the conclusion of the BVG Theorem (that a beginning is required of any expanding universe) as follows:

> Suppose, for example, that [a] space traveler has just zoomed by the earth at the speed of 100,000 kilometers per second and is now headed toward a distant galaxy, about a billion light years away. That galaxy is moving away from us at a speed of 20,000 kilometers per second, so when the space traveler catches up with it, the observers there will see him moving at 80,000 kilometers per second. If the velocity of the space traveler relative to the spectators gets smaller and smaller into the future, then it follows that his velocity should get larger and larger as we follow his history into the past. In the limit, his velocity should get arbitrarily close to the speed of light.[1]

This means that the relative velocity of the space traveler will appear to be smaller in the future—and larger in the past— than it is now. Remember, the rate

[1] Vilenkin 2006 p. 173. Alan Guth, at the University of California Santa Barbara's Kavli Institute, noted: "If we follow the observer backwards in an expanding universe, she speeds up. But the calculation shows that if $H_{average} > 0$ in the past, then she will reach the speed of light in a finite proper time."

at which the universe is expanding is increasing over time (see point four above).

In the future, when that galaxy is moving away from us at a speed of 30,000 kilometers per second (kps), the observers there will see him moving at 70,000 kps. So, in the past, when that galaxy was moving away from us at only 10,000 kps, the relative velocity of the space traveler would have been 90,000 kps.[1] Notice that the further we go back into the past, the faster relative velocities were. At some point we will reach the maximum velocity attainable for physical energy. In our universe, this maximum velocity is the speed of light -- but in another hypothetical universe it could be different. Once the traveler reaches the maximum velocity (speed of light in our universe), time would have reached its beginning, because, if time continued further past, he would be traveling faster than the speed of light, and this is impossible.

8) **Summary of evidence for a beginning from Alexander Vilenkin (at Hawking's birthday party).**

A brief summary of **Segment 2** of ***The Reason*** can be found in *New Scientist* issue 2984 (January 11, 2012) in which Alexander Vilenkin responds to Stephen Hawking's recent contentions about the eternity of past time.

Recall that Alexander Vilenkin is a very esteemed physicist and professor at Tuft's University in Boston. He was one of the principle discoverers of the BVG Theorem along with Arvin Borde (at UCSB) and Alan Guth (at MIT). In January of 2012, Vilenkin went to Stephen Hawking's 70th birthday celebration (on the state of the cosmos) to give people the whole story about what physics currently says about a beginning and creation. *New Scientist* reporter Lisa Grossman describes Alexander Vilenkin's correction of Hawking's omissions (about the evidence for creation from physics) as "the worst birthday present ever."

Basically, Vilenkin implied that Hawking left out all the evidence for a beginning of the universe from space-time geometry proofs (such as the BVG Theorem) and from entropy (the second law of thermodynamics). The following is a brief summary of some of Vilenkin's paper which summarizes **Segment 2**.

a. **Vilenkin's general assessment:**
"The hope of an eternal universe is fading, and may now be dead."

b. **Vilenkin's assessment of eternal inflation (and its multiverse) being infinite into the past:**
"You can't construct a space-time with the property [of eternal

[1] Every universe or multiverse must have a maximum velocity, because without it physical energy could travel at an infinite speed and be everywhere in the universe simultaneously. If that was the case, then velocity would be meaningless and everything would be everywhere. This would mean, for example, that protons and electrons are in the same place at the same time, which is a contradiction. Furthermore, there would be no laws of physics.

inflation or a multiverse that extends into the infinite past]. It turns out that the constant has a lower limit that prevents inflation in both time directions. It can't possibly be eternal in the past; there must be some kind of boundary."

c. **The evidence for a beginning of the universe (including a bouncing universe) in the *New Scientist*:**
"[The law of entropy requires that] disorder increases with time. So following each cycle, [in a hypothetical bouncing universe] the universe must get more and more disordered. But if there has already been an infinite number of cycles, the universe we inhabit now should be in a state of maximum disorder. Such a universe would be uniformly lukewarm and featureless, and definitely lacking such complicated beings as stars, planets and physicists—nothing like the one we see around us."

d. **Vilenkin's final conclusion:**
"All the evidence we have says that the universe had a beginning."

9. Conclusion

Since every possible universe we can imagine requires expansion, they all must have a beginning. Vilenkin's conclusion is that "past-eternal inflation without a beginning is impossible." And, as mentioned in **Segment 1**, a beginning of the universe requires that something other than it—like God—would have had to bring it into existence.

Handout 2a—Video Review and Discussion

Is There Any Evidence for a Creator in the Universe?

Opening Prayer

Father, you give us the ability to learn about the world that you created. Through our senses we can experience the things you have made. Through our minds we can think about how these things came to be and why they exist. May we be willing to take a closer look at what you have made and yearn to know more about your creation, and through them, know you. Amen.

Opening Reflection and Sharing

What do you think: *Is there any evidence for a creator in the universe? Do you think God's existence can be proven, or at least substantiated, by science?*

Review Questions

1. What does Tyler say science can do to any proof for God's existence that is developed from philosophy, and what evidence does he use to support his claim?

2. What does it mean that "before the universe existed, it was nothing?"

3. What can we learn from philosophy about what is required for things to exist?

4. Even though his Theory of Relativity had predicted that the universe was expanding, why didn't Einstein want to believe it?

5. What were the contributions of: a) Fr. Georges Lemaitre, and b) Edwin Hubbl, in convincing Albert Einstein that the universe was expanding and had a beginning?

Name: **Date:** **Period:**

6. What is some of the other evidence for the Big Bang that science has discovered?

7. Even though most physicists believe that the universe started with the Big Bang, why don't all of them?

8. What do these other physicists think are better explanations for the existence of our universe, and what evidence do they use to support this?

9. How do the beliefs of these scientists in speculative theories about the origin of the universe compare to Einstein's belief in the steady state theory?

10. What is the BVG Theorem, and what does it suggest about these speculative theories of the universe?

11. What is the significance of Dr. Vilenkin's statement, "Inflation without a beginning is impossible?"

Class Discussion Questions

1. What do you think of the claim that philosophy can prove there must be an eternally-existing being outside of the universe? Why?

2. Why do you think some scientists reject the evidence for the Big Bang, but accept the speculative competing theories?

3. What is the real significance of the BVG Theorem to science and faith?

4. Have you ever been biased as some modern physicists are, not accepting evidence that is verifiable while holding onto beliefs that are not? What did it take (or would it take) for you to finally accept the evidence in the way Einstein did?

Closing Prayer

Dear God, through science, you have given us the ability to learn that our universe had a beginning and that it could not have made itself, meaning that it was created. May we always be open to truth, whether it comes from science or philosophy, but most of all because it comes from you. Amen.

Handout 2b

What Caused Aristotle to Consider an Uncaused Cause?

Directions: Read the following, and complete the activity below.

Think about this: Aristotle (382-322 B.C.) argued that everything that begins to exist must have a cause and that, ultimately, there must be a *first*, or *uncaused*, cause. Aristotle reasoned that this first cause was the creator of the universe. How did he do this? He considered what would happen if you did not have a first cause (which would have to be uncaused). What would be the case if you didn't have a first cause that would be uncaused? You would have to have an endless regression of causes, one causing the next, causing the next, causing the next forever.

But he showed that even with this infinite regression, none of them could be real unless at least one of them had been uncaused. Why? Because something can't come from nothing.[1]

In a similar logical proof for God's existence, Aristotle argues that

[1] An example: Imagine an infinite regression of causes going backwards forever. Now notice that each one of these causes is actually nothing until it is caused by something. No matter how far back you go, all the causes are nothing because all of their causes are nothing without something real -- already existing -- to cause them. So Aristotle concluded this "something real which already exists" must be something which does not need a cause in order to be real. This is what he called his "uncaused cause" which he shows must be the first cause -- the creator.

Name: **Date:** **Period:**

there must be an unmoved mover. Like a series of dominos that cause one another to fall, there had to be something other than the dominos themselves causing them to start falling, such as a gentle push on the first one.

Assignment: Think of an everyday example of either of these "proofs"—uncaused cause or unmoved mover—and describe or illustrate it below. Be ready to share it with the class.

Handout 2c

Vilenkin's Response to Hawking: The Worst Birthday Present Ever

Students may wonder why some physicists such as Stephen Hawking or Richard Dawkins do not talk about the evidence for creation from space-time geometry proofs and entropy. Haven't these scientists heard of this evidence? It is virtually unthinkable that they have not heard about these well-known proofs. So what could be the explanation for omitting this evidence entirely? It seems that these supposed fair-minded intellectuals have intentionally omitted (or hidden) important evidence which conflicts with their position. Students must ask themselves two questions:

1. **How rigorous and objective could the opinions of Hawking and Dawkins be if they have intentionally omitted virtually every piece of evidence that directly contradicts their opinion?**

2. **Could such opinions be in any way considered correct?**

The answers should be self-evident.

This problem of intentionally omitting critical evidence came to a head in January 2012 at Stephen Hawking's 70th birthday celebration titled, "The State of the Universe." In a prerecorded interview before the conference, Hawking admitted:

> *A point of creation would be a place where science broke down. One would have to appeal to religion and the hand of God.*

Clearly, Hawking knows the implication of a beginning of the universe (or multiverse) addressed in **Segments 1 & 2** of ***The Reason***, and his desire to avoid "religion and the hand of God," may explain his omission of critical evidence.

Apparently, these glaring omissions provoked the esteemed physicist from Tufts University, Alexander Vilenkin, to call attention to the missing evidence. As Lisa Grossman from the *New Scientist*[1] wrote, "It was the worst birthday present ever."

Essentially, Vilenkin took the opportunity to present the evidence for creation of the universe from both space-time geometry proofs and entropy (which Hawking seems to have intentionally omitted).

The following excerpts from Grossman's article[2] closely parallel the evidence presented in this segment:

1. Vilenkin on Space-time Geometry Proofs (BVG Theorem):

   "In 2003, a team including Vilenkin and Guth considered what eternal inflation [a multiverse with many bubble universes] would mean for the Hubble constant... They found that the equations didn't work. 'You can't construct a space-time with this property,' says Vilenkin. It turns out that the constant has a lower limit that prevents inflation in both time directions. It can't possibly be eternal in the past. There must be some kind of boundary."

2. Entropy and Cyclic Universes

   Recall that the cyclic or bouncing universe is hypothesized to expand and contract indefinitely, but the law of entropy militates against this. As Grossman and Vilenkin note, "[The law of entropy holds that] disorder increases with time. So following each cycle, the universe must get more and more disordered. But if there has already been an infinite number of cycles, the universe we inhabit now should be in a state of maximum disorder… (such a universe would be dead), nothing like the one we see around us. One way around that is to propose that the universe just gets bigger with every cycle. Then the amount of disorder per volume doesn't increase, so needn't reach the maximum. But Vilenkin found that this scenario falls prey to the same mathematical argument as eternal inflation: if your universe keeps getting bigger, it must have started somewhere."

3. An Eternally Static Universe

   Vilenkin also disproves the possibility of an eternally static universe (prior to the Big Bang) in a proof also described in the *New Scientist* article.

4. Vilenkin's Final Conclusion:

   "All the evidence we have says that the universe had a beginning."

---

[1] Grossman, Lisa. "Why physicists can't avoid a creation event." *New Scientist* 11 Jan 2012, Issue 2847
[2] Ibid.

Assignment: Reflect on the following questions, either in small groups or individually.

1. If Vilenkin's assertions are true, how does this affect the debate between agnosticism and theism in physics?

2. If you were Vilenkin, how would you have handled this situation or one similar to it, where you strongly believe important information had been omitted, perhaps deliberately? Would you have gone to the birthday party to correct the omissions?

3. Is it enough to have *accurate* facts if you do not have *all* the facts when trying to discern the truth about the universe or God?

4. Would you give equal credibility to a scientist who omits evidence vs. the one who does not – if they are equally competent?

Name: **Date:** **Period:**

Notes

Name: Date: Period:

The Reason: **Segment 2 Quiz**

Is there any Evidence for a Creator in the Universe?

Matching

Match the name of the scientist in **Column A** with their contribution in **Column B**.

Column A

_____ 1. Albert Einstein

_____ 2. Alan Guth, Alexander Vilenkin and Arvin Borde

_____ 3. Edwin Hubble

_____ 4. Fr. Georges Lemaitre

Column B

a. Formulated the Big Bang Theory.

b. Contributed to the BVG Theorem.

c. Developed the general theory of relativity and the super-structure of modern cosmology.

d. Found evidence that the universe was expand

Modified True or False

If the answer is true, mark true, but if it is false, mark false and re-write the sentence to be a true statement.

Example:

A. ***The Reason*** is a video series that attempts to demonstrate that science and faith are incompatible.

Answer:

False A. ***The Reason*** is a video series that attempts to demonstrate that science and faith are ~~incompatible~~. compatible.

_____ 1. Philosophy can demonstrate that there has to be an eternally-existing being that always existed or nothing else could exist.

______ 2. While the evidence for the Big Bang is only speculative, there is much evidence for competing theories such as the bouncing universe or the multiverse.

______ 3. All scientists are always unbiased in evaluating scientific evidence.

______ 4. The BVG Theorem has only one condition for proving a universe would have a beginning, that it has an average rate of expansion greater than zero.

______ 5. The significance of the BVG Theorem is that any possible universe we can imagine would be an expanding universe and have to have a beginning.

______ 6. This segment of ***The Reason*** concludes that Tyler was right—science can disprove any proof for the existence of God from philosophy.

Meet the Scientists

Albert Einstein

Dr. Albert Einstein (b. 1879 in Germany, d. 1955) is perhaps the most well-known cosmologist and physicist of the 20th century. He is considered the father of modern physics for revolutionizing the field with his 1916 theory of General Relativity. For this and other achievements he received a Nobel Prize in physics in 1921. Even though his theory predicted an expanding universe, Einstein added a cosmological constant to it supporting his belief in an eternal and, therefore, a non-expanding, universe. When shown the incompatibility between the mathematics of his theory and its conclusions, Einstein removed the cosmological constant, calling it the greatest blunder of his career. Einstein became a U.S. citizen in 1940, teaching physics at the Institute for Advanced Study at Princeton until his death.

Georges Lemaitre

Fr. Georges Lemaitre (b. 1894 in Belgium, d. 1966) was a priest, astronomer, and professor of physics at the Catholic University of Louvain. In 1927, Fr. Lemaitre published an article presenting what became known as Hubble's Law, showing that the universe is expanding. Although calling the origin of the expansion of the universe the primeval atom, he eventually became known as the Father of the Big Bang Theory (The term Big Bang was sarcastically given later to Lemaitre's theory by Sir Fred Hoyle, but the name stuck.) Fr. Lemaitre became the head of the Pontifical Academy of Sciences from 1960 until his death.

Meet the Scientists

Edwin Hubble

Dr. Edwin Hubble (b. 1889 in Missouri, d. 1953) earned his Ph.D. in astronomy from the University of Chicago in 1917 and in 1917 began working at the Mount Wilson Observatory near Pasadena, California. Observations between 1922 and 1923 led to his discovery that the universe was much larger than the Milky Way Galaxy. In 1929, using observations of the red-shift of distant galaxies, Hubble formulated what is known as Hubble's Law describing the expansion rate of the universe. This provided empirical confirmation of Einstein's Theory of Relativity. He remained on the staff of the Wilson Observatory until his death.

Arvin Borde

Dr. Arvin Borde earned his Ph.D. from the State University of New York at Stony Brook in 1982. Currently, he is the Senior Professor in the Department of Mathematics at Long Island University. He was recently the KITP Scholar and a General Member of the Kavli Institute of Theoretical Physics at the University of California at Santa Barbara. In 2003, Borde collaborated with Drs. Vilenkin and Guth on a study and report titled *Inflationary Spacetimes are Incomplete in Past Directions*, published in *Physical Review Letters*. This is the basis of the BVG Theorem.

Meet the Scientists

Alexander Vilenkin

Dr. Alexander Vilenkin emigrated from the former Soviet Union in 1976 after being blacklisted for not cooperating with the KGB. He earned his Ph.D. in physics from Buffalo University and is the Director of the Institute of Cosmology at Tufts University. Vilenkin collaborated on the BVG Theorem. In his 2006 book, *Many Worlds in One,* he wrote:

> *It is said that an argument is what convinces reasonable men and a proof is what it takes to convince even an unreasonable man. With the proof now in place, cosmologists can no longer hide behind the possibility of a past-eternal universe. There is no escape: they have to face the problem of a cosmic beginning.*

Alan Guth

Dr. Alan Guth (b. 1947 in New Jersey) earned his Ph.D. in physics from MIT in 1972, where he is currently a professor. He developed the idea of cosmic inflation—the theorized rapid expansion of the universe during the first milliseconds of its existence—in 1979. He collaborated on the BVG Theorem in 2003. Guth was awarded the Cosmology Prize of the Peter Gruber Foundation in 2004, and the Isaac Newton Medal in Physics from England's Institute of Physics in 2009.

Notes

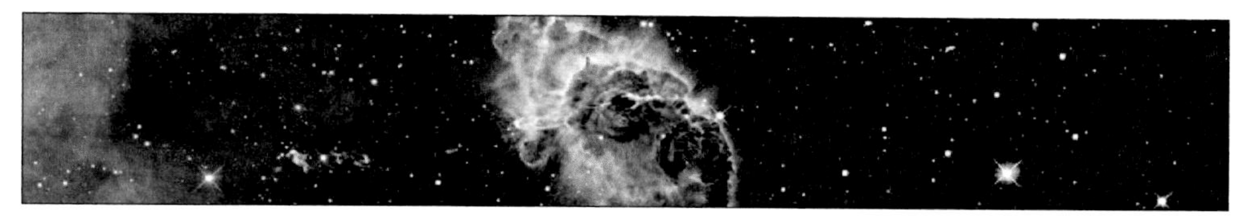

Segment 3

Is the Universe Random and Meaningless?

Objectives

Students will:

1. Learn examples of the conditions and constants of the universe,

2. Understand that the actual values of these conditions and constants are extremely improbable and that they are just what is needed for advanced life to exist, and

3. Explain why many physicists call this improbability fine-tuning and consider it very likely that a super-intelligent creator planned it that way.

Overview

Once again, we meet Joe visiting Alana and Dan to share how his discussions with Tyler, his roommate, are going and ask for help in answering Tyler's objections. Joe's new knowledge is influencing Tyler, who now accepts that, even if there was something existing prior to our universe, it still would have to have a beginning, and, therefore, a creator. Tyler, however, is still holding on to the idea that humans are just an accident of evolution and that the creator did not plan our existence or care about us. Alana and Dan share good news with Joe. The conditions and constants of the universe demonstrate that the creator is extremely intelligent and it is extremely improbable that life could have arisen by chance.

Scientific Summary of Segment 3

Is the Universe Random and Meaningless?

1. **Other than the space-time geometry proofs and the fact that our universe is not completely run down from entropy as described in Segment 2, is there any other scientific evidence useful to show God's existence?**

   Yes, due to the highly improbable conditions (such as low entropy) and constants (such as the speed of light) of the universe which allowed for life to emerge and develop (which physicists refer to as fine-tuning), there is evidence that a super-intellect designed it. This means that the universe has the very specific values required for life to exist.

   However, there is no reason for these values to be what they are; they could be almost anything. Since these values are precisely what they need to be for life to exist, and this is exceedingly, exceedingly improbable, it is unlikely that they occured by pure chance.

   This kind of universe is also known as an anthropic (capable of supporting life forms) universe with anthropic coincidences (conditions and constants.)

2. **What is an anthropic condition, and how is the low entropy of our universe at the Big Bang an example of an anthropic condition?**

   Anthropic conditions are those which must be present in the universe in order for any life form to emerge.

   Entropy refers to the irreversible process of the energy and matter in our universe moving from an ordered and usable state (low entropy) to a disordered and unusable state (high entropy). When the universe comes close to a state of maximum (high) entropy, it is run down and can no longer do anything. This is sometimes called a dead universe.

   Without there being any known physical reason requiring it, our universe had very low entropy at the Big Bang, making it well suited for the emergence and development of life.

   What are the odds of our universe having low entropy at the Big Bang? Physicist Roger Penrose calculated the odds against it as an astounding $10^{10^{123}}$ to one! This is a double exponent, and if it were written out in normal exponential notation, the exponent would have a one followed by 123 zeros. If we wrote this number out in a non-exponential way, and every zero was 10-point type, the number would fill up our galaxy. This number is similar to the odds of a monkey randomly tapping keys on a typewriter producing the complete works of William

Shakespeare. In other words, exceedingly, exceedingly improbable (many physicists call this "virtually impossible").

No one can seriously consider either of these to be purely random occurrences. Since there is no *natural* explanation (based on evidence) for the extreme low entropy of the universe at the Big Bang, physicists have made recourse to one of two *metaphysical* explanations. One explanation is a multiverse which would allow for trillions upon trillions of different levels of entropy in each of its countless bubble-universes. However, this hypothesis has several problems. First, there is no physical evidence for this hypothesis, and as we saw in **Segment 2,** multiverses must have a beginning which means there cannot be an infinity of bubble universes. Furthermore, all known multiverse hypotheses must have considerable (highly improbable) fine-tuning in their initial conditions. It seems then that current multiverse hypotheses do not solve the problem of highly improbable anthropic conditions (all it has done is move the problem back one step—from our universe to the multiverse).

So what is left? Currently the most reasonable and responsible explanation seems to be what Sir Fred Hoyle called a super intellect which selects the values of our universe's conditions and constants, like the creator we encountered in **Segment 2.**

3. **What is an anthropic constant, and what are some anthropic constants of the universe?**

A constant is a numerical value which determines the specific properties and outcomes of the equations of physics (like the speed of light) and controls the laws of nature. Anthropic constants are those values that are desirable for the emergence and development of life. There are about 20 fundamental constants in our universe, such as:

a. **The speed of light constant:** (186,200 miles per second or 300,000 km per second). This determines the invariant speed of light in all reference frames and the highest attainable velocity of energetic systems in the universe. (Remember our space traveler in **Segment 2**? Nothing can travel faster than the speed of light in our universe.)

b. **All four forces in our universe have constants which determine their properties and outcomes:**

(1) The gravitational constant ($G = 6.67 \times 10^{-11}$)

(2) The strong nuclear force coupling constant ($g_s = 15$),

(3) The weak force constant ($g_w = 1.43 \times 10^{-62}$)

(4) The electromagnetic force. The electromagnetic force has three constants associated with it:

- The mass of a proton ($mp = 1.67 \times 10^{-27}$ kg)
- The mass of an electron ($me = 9.11 \times 10^{-31}$ kg)
- The electromagnetic charge which pertains oppositely to both protons and electrons ($e = 1.6 \times 10^{-19}$ coulombs).

c. **There are about thirteen other constants, some of which you have probably heard of:**

(1) Hubble's Constant ($H = 2 \times 10^{-18}$ SI units), which governs the rate of expansion of the universe (which we already saw in our last segment)

(2) Planck's Constant

(3) The Cosmological Constant

(4) The Planck minimum length of space

(5) The Planck minimum length of time

4. **Why are the *anthropic* constants (which allow for life forms) of the universe not thought to have a natural explanation?**

Because they are exceedingly, exceedingly improbable and their occurrence by pure chance is highly unlikely. The following four examples, out of many, many more, are commonly used in many physics books, including Stephen Hawking's *The Grand Design*:

a. **An ordered universal expansion allowing for life.**
If the gravitational constant (G) or the weak force constant (g_w) varied from their values by only one part in 10^{50} higher or lower (.000000000000000000000000000000000000000000000000000000001 – a very small fraction), then the universe would have:

(1) Continually exploded in its expansion (which would have prohibited the emergence of any life form), or

(2) Collapsed into a black hole (which likewise would have prohibited the emergence of any life form.)

Are we to believe that these incredibly improbable, but necessary, values of the gravitational and weak force constants really occurred by pure chance?

b. The elements of the Periodic Table.
If the strong nuclear force coupling constant (g_s) varied from its value by only 2% more, then there would be no hydrogen in the universe (no nuclear fuel, water, etc., prohibiting the emergence of any life form.)

If it were only 2% less than its value, there would be no element heavier than hydrogen, such as carbon, the building block of life (which would likewise prohibit the emergence of any life form).

Are we to believe that the strong nuclear force coupling constant received its value within this very narrow window by pure chance?

c. The emergence of stars allowing for the emergence of life.
The stars in our universe are at the point of convective instability—that is, they are sitting in the exceedingly narrow range between:

(1) Blue giants (exploding stars which are far too hot to allow for the emergence of life forms), and

(2) Red dwarfs (which are too weak to supply the radiation necessary for the emergence and development of life forms over time).

If the gravitational constant, electromagnetism, or the mass of the proton in relation to the mass of the electron had varied ever so slightly from their values, then all the stars in the universe would have been either blue giants or red dwarfs.

Are we to believe that this exceedingly improbable coincidence of the values of these four constants necessary for the emergence of any life form occurred by pure chance?

d. The emergence and abundance of carbon in our universe.
This coincidence convinced then atheist Sir Fred Hoyle of the existence of a "Supercalculating Intellect" as the source of the universe. The resonance levels (which control the "stickiness" and bonding properties of atoms) of helium, beryllium, oxygen, and carbon have to fit within a very narrow range of possible values in order for there to be an abundance of carbon (the building block of life.) These values are so improbable that Hoyle actually compared their emergence in the universe by pure chance to those of a "tornado sweeping through a junkyard and constructing a Boeing 747 ready for flight."

Are we to believe that this highly improbable coincidence of resonance levels occurred by pure chance?

5. **Why do the anthropic coincidences of the universe seem to require a supernatural explanation?**

   There is general agreement among both theistic (those who believe in a supernatural creator) and agnostic (those who are not sure) physicists that pure chance is not an adequate explanation for the above mentioned anthropic coincidences. The improbability is simply too overwhelmingly high against them. This has left the physics community with only two possible explanations:

   a. **One, the multiverse speculation which allows for a non-intelligent, random occurrence of anthropic coincidences through trillions upon trillions of hypothetical universes; or**

   b. **Two, an intelligent cause which selected the values perfectly.**

   Although there is no direct physical evidence of either a multiverse or of a supernatural designer or intelligence, they represent the best explanations we have for the anthropic coincidences which cannot be explained by pure chance.

6. **What factors should be considered when making a decision about which of these two explanations—the multiverse speculation or an intelligent cause—is more likely?** There are three factors:

   a. **Current multiverse hypotheses, such as string theory, have serious problems and have been criticized by several leading physicists such as Michael Dine.**

   b. **Current multiverse theories require considerable (improbable) fine-tuning in their initial conditions and constants, meaning that they do not answer the problem of highly improbable anthropic coincidences (they only move the problem back one step - from our universe to a hypothetical multiverse).**

   c. **Because of the evidence for a beginning of the universe (and its implication—a transcendent creator), the idea of an intelligent cause is not a blind leap to the supernatural, but rather an attributing of the quality of intelligence to a supernatural cause for which other evidence already exists (such as the BVG Theorem and the current low entropy of our universe—see Segment 2).**

7. **Conclusion: So, a supernatural, supercalculating intellect should be considered a reasonable and responsible explanation for the anthropic coincidences in our universe.** It may well be the only reasonable and responsible explanation if current multiverse hypotheses continue to have serious problems.

Handout 3a—Video Review and Discussion

Is the Universe Random and Meaningless?

Opening Prayer

Lord, we live in a universe that is so immense we can't imagine all it contains, yet so intricate we can't fully understand how or why familiar things work the way they do. Open our eyes to see what you have done, open our minds to comprehend the mastery you have over creation, but most importantly, open our hearts to appreciate what you have done for us. Amen.

Opening Reflection and Sharing

What do you think: *Is the universe random and meaningless?*

Review Questions

1. While Tyler seems to accept the science behind the BVG Theorem, that any universe that's expanding has to have a beginning, what does he say he's still sure about?

2. What types of scientific evidence exist to suggest we have an extremely intelligent creator?

3. What is it that leads physicists to think that the conditions and constants of the universe suggest an extremely intelligent creator?

4. How is the very low level of entropy of the universe an example of one of these conditions?

Name: **Date:** **Period:**

5. What would have happened if the universe had a high level of entropy?

6. How is the low level of entropy of our universe evidence for an intelligent creator?

7. Besides an intelligent creator, what other explanations are being discussed by physicists for the fact that our universe has the very low entropy needed for life to exist?

8. If other universes do exist and have different levels of entropy, would that be an argument against there being a creator?

9. What is the sports car engine used to illustrate?

10. What is meant by the phrase, "constants of the universe"?

11. What are some of the laws of nature that describe these constants?

12. How are the constants evidence of an intelligent creator?

13. How did Sir Fred Hoyle, an atheist, react after discovering that the existence of carbon, the building block of life, was so incredibly unlikely that it didn't make sense to think it was an accident?

14. What are the two reasons given so far as evidence to support the existence of a powerful, intelligent creator?

Class Discussion Questions

1. After watching this video, what do you think of Tyler's claim that the universe is random and meaningless?

2. Of all the examples given in the video about the conditions and constants of the universe being fine-tuned, which made the point easiest for you to understand? Why?

3. Why do you think some scientists reject the evidence for the fine-tuning of the universe but accept the speculative competing theories?

4. What is the real significance of the evidence of the fine-tuning of the universe to science and faith?

Closing Prayer

Heavenly father, we have learned through our study of the universe that you are very powerful and intelligent; that you created and planned the universe so that life could exist, but not just any life… us. The universe is a sign of your great love for us. It was created to give us a place to live and designed to sustain us. Thank you for your wonderful gifts of creation and life. Amen.

Name: **Date:** **Period:**

Notes

Handout 3b

Why do Physicists Say the Universe is Finely Tuned?

Directions: Go to <http://www.youtube.com/watch?v=guHodt-7Q7A&feature=related> and watch the video titled **The Finely Tuned Universe.**

Then complete the following questions:

Part One

1. What is it about the universe that surprises scientists?
2. Why are the fundamental forces and constants of the universe referred to as being finely tuned?
3. What would the impact be on complex life if any one of the fundamental constants were only slightly different in its value?
4. What conclusions are reached about the cause and purpose of the fine-tuning?

Part Two

In light of the fine-tuning evidence, explain this quotation from Sir Fred Hoyle:

> *"A common sense interpretation of the facts suggests that a super intellect monkeyed with physics, as well as with chemistry and biology, and that there are no blind forces worth speaking about in nature. The numbers one calculates from these facts seem to me so overwhelming as to put this conclusion beyond question."*

Name: **Date:** **Period:**

Notes

Name: Date: Period:

The Reason: Segment 3 Quiz

Is the Universe Random and Meaningless?

Modified True or False

If the answer is true, mark true, but if it is false, mark false and re-write the sentence to be a true statement.

Example:

_____ A. ***The Reason*** is a video series that attempts to demonstrate that science and faith are incompatible.

Answer:

False A. ***The Reason*** is a video series that attempts to demonstrate that science and faith are ~~incompatible~~. compatible.

_____ 1. Scientific evidence now exists that proves the universe is random and meaningless.

_____ 2. Entropy is an example of a condition of the universe.

_____ 3. We live in a universe of extremely high entropy.

_____ 4. Roger Penrose calculated that the odds against our universe being low entropy were 1 in 50.

_____ 5. Some physicists choose to believe that it's more likely there are trillions upon trillions of universes, rather than there being one intelligent creator.

_____ 6. If evidence for other universes were discovered, a creator of the multiverse (in which they would have to be contained) would still be necessary.

_____ 7. The speed of light is an example of a constant of the universe.

_____ 8. There are about eight constants that govern everything in our universe.

______ 9. Physicists are baffled at the value of the constants because they are exactly what they need to be to support life (yet they do not have to be that way).

______ 10. After researching carbon, the building block of life, Sir Fred Hoyle, an atheist, admitted that it was common sense to believe in a super-intellect outside of our universe.

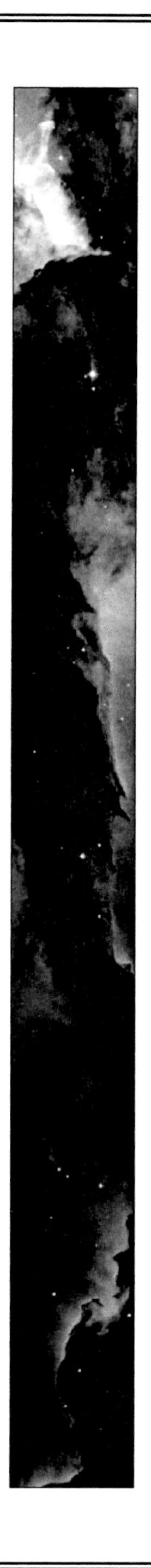

Meet the Scientists

Roger Penrose

Dr. Roger Penrose (b. 1931 in England) earned his Ph.D. in Physics from Cambridge in 1958. An atheist most of his life, his studies on the low entropy at the beginning of the universe led him to consider an intelligent cause rather than random chance as an explanation. In Stephen Hawking's video: *A Brief History of Time*, Penrose said:

> *I think I would say that the universe has a purpose, it's not somehow just there by chance... some people, I think, take the view that the universe is just there and it runs along – it's a bit like it just sort of computes, and we happen somehow by accident to find ourselves in this thing. But I don't think that's a very fruitful or helpful way of looking at the universe, I think that there is something much deeper about it.*

Paul Davies

Dr. Paul Davies (b. 1946 in England) completed his Ph.D. in Physics from University College London in 1970. Later, he studied under astronomer Sir Fred Hoyle at Cambridge. Davies has written extensively on scientific and philosophical issues, including the relationship between science and faith, for which he has taken criticism from some physicists and astronomers. In reaction to the criticism he has said:

> *I was dismayed at how many of my detractors completely misunderstood what I had written. Indeed, their responses bore the hallmarks of a superficial knee-jerk reaction to the sight of the words 'science' and 'faith' juxtaposed.*

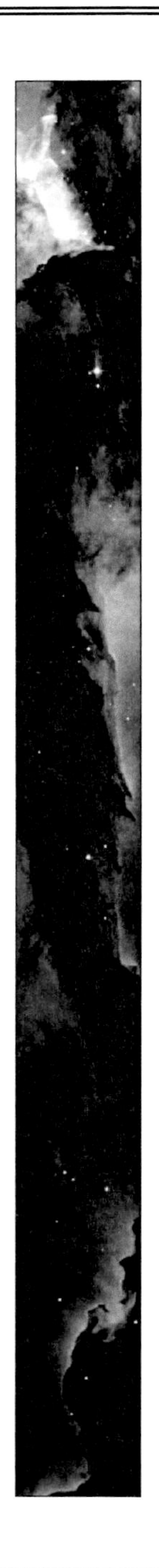

Meet the Scientists

Fred Hoyle

Sir Fred Hoyle (b. 1915 in England, d. 2001) is known for his rejection of the Big Bang Theory, a term which he coined to sarcastically describe the conclusion that, because of the expansion of the universe, it would have had to begin at a single point, at a finite time in the past. But, because he was open to having his steady state theory challenged, he later accepted the Big Bang concept. Studies on the characteristics of carbon led him to conclude that:

> *A common sense interpretation of the facts suggests that a superintellect has monkeyed with physics, as well as with chemistry and biology, and that there are no blind forces worth speaking about in nature. The numbers one calculates from the facts seem to me so overwhelming as to put this conclusion almost beyond question.*

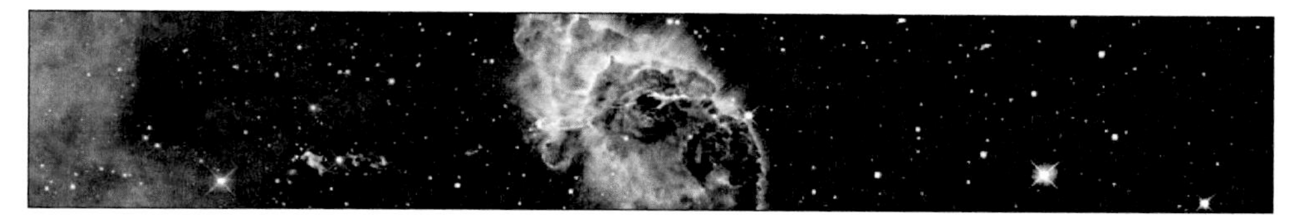

Segment 4

Does the Bible Conflict with Science?

Objectives

Students will learn:

1. That the Bible is a book of theology and not of science, and, therefore, can't conflict with science;
2. That philosophy and science are limited to describing God as transcendent, powerful, and intelligent, but it is quite difficult to go further with these methods; and
3. That if we are created with a desire for unconditional love, and come from God, God must possess that quality also.

Overview

Joe is back visiting with Alana and Dan to ask more questions and report on his progress with his roommate, Tyler. By now, Tyler admits that the universe had to be created and is even willing to consider that its conditions may be unlikely. What he is not ready to admit, however, is that the creation of the universe has anything to do with the God of the Bible, thinking its stories to be nothing more than ancient myth. Alana and Dan explain that the Bible can't conflict with science because the Bible is not about science. The Bible is about theology, and the truths that science and the Bible reveal complement each other. Not only is God very powerful and super intelligent, he also loves us unconditionally.

Philosophic Summary of Segment 4
Does the Bible Conflict with Science?

1. **Is there a contradiction between the scientific and Biblical accounts of creation?**

   No, because the Biblical creation accounts were never intended to be scientific. They are theological. The Biblical author of Genesis (who lived before 500 B.C.) could never have understood science as we know it. God would not have inspired the Biblical author with an explanation of creation containing complex math and science which would have been completely unintelligible to him and his audience.

2. **What was God doing when he inspired the Biblical author?**

   God was giving the Biblical author theological solutions to his theological problems in the understanding of his time and day. For example, myths like the Epic of Gilgamesh (which competed with Jewish theology), taught:

   a. **There were many gods;**
   b. **Natural objects, such as the sun, moon, stars, and sea were gods; and**
   c. **These gods were often unjust and fashioned the world in a way that was filled with both good and evil.**

   So, the Biblical author had to correct these theological errors before they became confusing to the Jewish people. They needed their own creation epic to counter the errors in the rival epics, and so the Biblical author was inspired to write an epic where:

   a. **There was one God;**
   b. **This God created everything else such as the sun, the moon, the stars, and the seas;**
   c. **This God was not capricious or unjust, and he certainly did not toy around with human beings;**
   d. **He created a world which was fundamentally good (which he recognized to be "good"); and**
   e. **He made human beings in his own image – having a divine dignity.**

   To read the Bible looking for a scientific explanation of creation is a misunderstanding of what divine inspiration is and how it works. Divine inspiration is not a dictation of scientific truth, but rather an inspiration of theological truths the author and his audience could understand.

3. **Why was there a need to inspire Biblical authors with theological truths?**
   As we already saw in **Segment 2** and **Segment 3**, there is abundant evidence from physics and philosophy that a creator exists who is highly, highly intelligent. However, science and philosophy can give only limited evidence

about God. So what can science and philosophy show or prove?

a. **There is a transcendent creator;**
b. **This creator has enough power to create the universe as a whole;**
c. **God is one, and is not subject to space, time, or other limitations; and**
d. **God is probably highly intelligent and perhaps even unrestricted in intelligence.**

So what questions can't physics and philosophy adequately answer? Some of the very basic ones are:

a. **Is God love?**
b. **Is God unconditional love?**
c. **Does God redeem suffering?**
d. **Does God answer prayers?**
e. **Does God guide us in our everyday lives?**
f. **Does God make good come out of evil?**

Philosophy and science can give us knowledge about the nature of a highly intelligent supernatural power, but they cannot be certain about what the super-intelligence's heart (emotions or feelings) is like. There are some philosophers and scientists who believe that God is purely rational (and has no emotions or feelings to speak of). This view was held by Aristotle in the Classical Period, by many deists in the 17th through 20th centuries (including several of founding fathers of the U.S.), and by many modern scientists (including Einstein).

4. How does Christianity differ from other religions in its claims about God?

While many of the world's religions have:

a. **A sense of God's involvement in creation and interaction with human beings, and**

b. **An implicit and/or explicit recognition of God's care and guidance and various explanations of human suffering,**

Christianity makes three extraordinary claims which seem to go beyond the revelation of most other religions:

a. **Love is the highest virtue and all other virtues are subordinate to it and find their purpose and end in it;**

b. **God is not only love, but unconditional love, which is unconditionally forgiving (like the father of the prodigal son) and unconditionally affectionate (as demonstrated in Jesus' name for God—**

Abba or "daddy") and unconditionally compassionate (like a divine "good Samaritan"); and

c. **God so loved the world that he sent his own son (his beloved one) to be among us in a radical act of empathy, care, and self-sacrificial love to be with us "peer to peer, face to face, and brother to brother and sister."** God is truly "Emmanuel—God with us."

If this revelation of Christianity is true, namely, that God loves us infinitely and unconditionally (even to the point of ultimate sacrifice), then we can be sure that:

a. **There is no ultimate tragedy in this world,**

b. **Even the most grief-filled suffering will be brought to redemption in eternal and unconditional love, and**

c. **Even the greatest sinner can be forgiven through an act of sincere repentance.**

These truths allow us to hope in the face of despair and give meaning and purpose for our lives. They also inspire us to build the Kingdom of eternal love by enkindling love and faith in Emmanuel—Jesus Christ—in the hearts of the people we touch.

5. **How can we know the revelation of Jesus Christ is really the truth?**

There is a great deal of historical evidence for Jesus Christ—his resurrection, the gift of his divine Spirit and his miracles. Much of this evidence has come to light from truly great scholars in the 1960s, (e.g. Joachim Jeremias) the 1970s and 1980s (e.g. Raymond Brown and Joseph Fitzmyer), and the 1990's through today (N.T. Wright and John P. Meier). But is there a way of getting to the reality of God with us? The following questions may present a path to the truth about Jesus as Emmanuel (God with us).

a. **What is the most positive and creative power or capacity within me?** There is only one human power always focused toward good, and, therefore, able to direct our intellect and creativity to their proper purpose. That power is love. Love's capacity for empathy and its ability to unite with others leads to a natural giving of self for the good of others and the whole human community.

b. **If love is the one power that always seeks the positive and we are made to find our purpose in life through love, could God (who created the universe in a way that would lead to intelligent, free, and loving beings) be without love?**
Consider this: if the creator does not love, why would he:

(1) Create human beings not only with the capacity for love, but to be fulfilled only when they love?

(2) Make love the actualization of all human powers and desires, and, therefore, of human nature?

If the creator does not love, then the creation of beings meant for love seems ridiculous. However, if the creator is love, then creating loving creatures and sharing his loving nature with them would be consistent with what (or perhaps better, "who") he is.

c. **Is my desire to love and to be loved conditional or unconditional?**
We appear to have a desire for perfect and unconditional love. Not only do we have the power to love, we have a sense of what perfect love is like. This sense of perfect love has the positive effect of encouraging us to pursue perfect love. A drawback of this is that we expect perfect love from other human beings, who are incapable of giving it to us.

d. **If my desire for love can only be ultimately satisfied by unconditional love, then could the creator of this desire be anything less than unconditional love?**
If we assume that the creator does not intend to frustrate our desire for unconditional love, then his creation of the desire implies his intention to fulfill it, which also implies the very presence of this quality within him. This would mean that the creator of the desire for unconditional love (as the only possible fulfillment of that desire) would have to be unconditional love.

e. **If the creator is unconditional love, would he want to enter into a relationship with us of intense empathy; that is, would he want to be Emmanuel (God with us)?**
Love is empathizing with others and entering into a unity with them so that doing good for them is just as easy, if not easier, than doing good for oneself. This kind of love has the humility, self-gift, deep affection, and care which would turn infinite power into infinite gentleness and cause an infinitely powerful being to enter into a restrictive condition (like human life) to empathize more fully with his loved ones. Being Emmanuel (God with us) would be typical of an unconditionally loving God.

f. **If it would be typical of the unconditionally loving God to want to be fully with us, then is Jesus the one?**
Consider the following:

(1) Jesus came to reveal that love is the highest commandment (and that this highest commandment reflects the heart of God);

(2) He revealed God's name (God's essence in Semitic culture) to be "Abba" (that is, affectionate and trustworthy father—literally, "daddy");

(3) His ultimate revelation of who God is came through two fundamental figures in his parables, the father of the prodigal son and the compassion of the good Samaritan;

(4) Jesus made it his business to associate with and even share fellowship with sinners (which caused great scandal among the Jewish religious authorities);

(5) He healed, cured, and exorcised those who were in need and earnestly requested his help;

(6) He gave himself up to a self-sacrificial death in what he considered to be an act of unconditional love; and

(7) He shared the divine spirit who reveals in our hearts that God is precisely who Jesus said he was – unconditional love for all human beings.

Consider further the:

(8) Considerable historical evidence for his Resurrection,

(9) Miracles, and

(10) Miraculous manifestation of the Spirit.

If we examine the nature of love, our ultimate purpose as loving, and Jesus' preaching about God as unconditional love, then it seems not only plausible but true that Jesus is Emmanuel— God with us.

6. Conclusion

The Bible is not explaining science but theology, and theology is necessary to complement the evidence of science and philosophy in order to reveal the heart of God. Christianity gives the ultimate revelation of God's heart by showing, through Jesus Christ, that God is unconditional love: i.e., unconditional forgiveness, affection, empathy, care, and compassion.

It is both reasonable and responsible to believe in Jesus' revelation. This is shown in the considerable historical evidence that he is the beloved son sent by the father in unconditional love to be with us, reveal himself to us, and redeem us, so that we might participate in his unconditional love for eternity.

Handout 4a—Video Review and Discussion

Does the Bible Conflict with Science?

Opening Prayer

Oh creator God, we have learned that there are many reasons from philosophy and much evidence from science to know that you are very powerful and super-intelligent. However, we live in a world that often ignores and disregards the reasons and evidence that support your creative work. Help us to be open to your presence in all things and see you at work in our lives wherever we are and in whatever we do. Amen.

Opening Reflection and Sharing

What do you think: *Does the Bible conflict with science?*

Review Questions

1. What is it that Tyler says no one could ever convince him of and why?

2. What is it about the age of the universe that seems to present a conflict between science and the Bible? Is there really such a conflict?

3. What was the ancient Greek philosopher Aristotle able to prove? What couldn't he prove?

4. What are philosophy and science able to tell us about God? What do we need theology for?

5. Because God created us with a desire for unconditional love, what does that tell us about him?

6. In light of God's unconditional love, why does it make sense to believe in Jesus?

Name: **Date:** **Period:**

7. Why is it incorrect to say that God created evil or that when people experience evil it comes from God?

8. What does it mean to say that the Biblical authors were inspired, and what does this mean for our understanding of the Bible?

9. How does this concept of inspiration apply to the different ages of the universe (as demonstrated by science and as found in the Bible)?

10. Why were myths important to ancient peoples, and why was it so important for the Biblical authors to provide a creation story?

11. What teachings from the Babylonian creation myths were the Biblical authors trying to correct?

12. How is Sir Arthur Eddington (one of the greatest astronomers and physicists) an example of a person having a correct understanding of the relationship between science and the Bible?

Class Discussion Questions

1. After watching this video, what do you think of Tyler's claim that anyone who believes in the Bible can't believe in science because they totally contradict each other?

2. What are some implications for our world today in properly understanding the Bible as a theology book and not as a science or history book?

3. What are some things we can know about God from the Bible that we couldn't know from philosophy and science alone?

4. Why is it important to understand God's unconditional love for us to realize the significance of Jesus' life and saving actions?

Closing Prayer

Lord, from philosophy and science we learn that you are all-powerful and all-knowing, but to know you are all-loving as well, you had to reveal that to us. We ask you to help us be open to your truth as revealed in the Bible. May we remember that the very presence of your son on the Earth, as your free gift of salvation, was a sign of your unconditional love for us. Amen.

Handout 4b

What's True About the Creation Story?

The diagram on the right depicts the universe as ancient peoples, including the Hebrews, would have understood it. The earth is at the center, flat, with a dome above it with floodgates to let water in. The sun, moon and stars were inside the dome to give light, and moved around the unmoving earth.

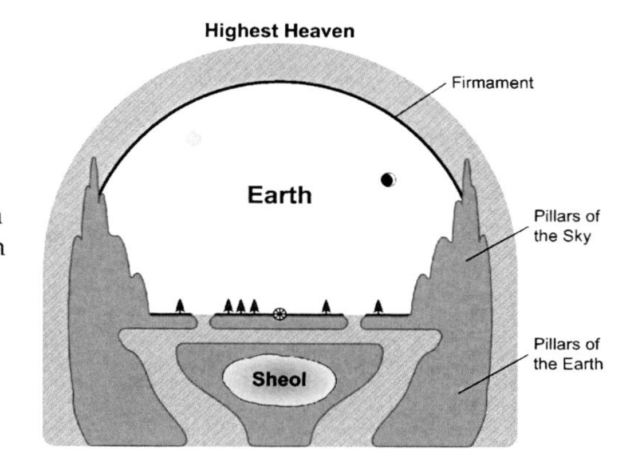

Directions: Read the first creation story of the Bible (Genesis 1:1 to 2:4) and answer the following questions in light of the diagram above.

1. Find a detail from each day of creation that describes this model of the universe:

   a. The first day

   b. The second day

   c. The third day

   d. The fourth day

   e. The fifth day

   f. The sixth day

2. Each of the six days has a three-part pattern. Complete this statement:

   Each day of creation begins with God: a)_________; then: b) _________; then God said: c)_________.

Name: **Date:** **Period:**

3. Look more closely at this creation story and you will notice another pattern: the first and fourth days, the second and fifth days, and the third and sixth days are related, but how?

   a. The first and the fourth days are both about:_________.
   b. The second and fifth days are both about:_________.
   c. The third and sixth days are both about:_________.

4. Keeping in mind what was said about inspiration in **Segment 4** of ***The Reason*** and the patterns of the first creation story in Genesis, what do you think are the theological truths of this creation account?

5. Why would it be incorrect to interpret this creation story scientifically or historically?

Notes

Name: Date: Period:

The Reason: Segment 4 Quiz

Does the Bible Conflict with Science?

Modified True or False

If the answer is true, mark true, but if it is false, mark false and re-write the sentence to be a true statement.

Example:

______ A. ***The Reason*** is a video series that attempts to demonstrate that science and faith are incompatible.

Answer:

False A. ***The Reason*** is a video series that attempts to demonstrate that science and faith are ~~incompatible~~. compatible.

______ 1. The Bible conflicts with science.

______ 2. The Bible teaches as fact that the universe is only six to ten thousand years old.

______ 3. Philosophy and science give us reasons and evidence to know that there is a very powerful and super-intelligent creator of the universe.

______ 4. We needed God himself, through the Bible and theology, to reveal that he is loving and cares for us.

______ 5. The fact that we are created to desire unconditional love tells us that God has unconditional love for us.

______ 6. Since God created everything, when we experience evil, we can say it comes from God.

______ 7. In the Bible, inspiration means that God dictated the exact words that the Biblical authors were to write down.

_____ 8. Because of the process of inspiration, we should not take specific details of the Bible literally without knowing if that is what the authors intended.

_____ 9. Ancient cultures used myth as a way to keep their people ignorant of the truth.

_____ 10. The authors of the Biblical creation stories used myth as a way to counter the false teachings their people were encountering in the Babylonian myths.

Meet the Scientists

Sir Arthur Eddington

Sir Arthur Eddington (b. 1882 in England, d. 1944) was an astrophysicist whose achievements included verification of stellar parallax—first predicted by Einstein's Theory of Relativity—during a total solar eclipse in 1919, and his studies on the natural limit of the luminosity of stars, for which the Eddington Limit is named. He also explained relativity to the English audience. As a writer he was willing to discuss the religious implications of physics and argued that new findings in physics allowed room for personal religious experience and free will.

Robert Jastrow

Dr. Robert Jastrow (b. 1925 in New York, d. 2008) received his Ph.D. in theoretical physics in 1948 and joined NASA when it formed in 1958, eventually becoming director of the Goddard Space Institute for Space Studies in 1961 until his retirement 20 years later. He was open to the creation of the universe at the Big Bang. He has said:

> *Science has proved that the universe exploded into being at a certain moment. It asks: 'What cause produced this effect? Who or what put the matter or energy into the universe?' And science cannot answer these questions.*

Notes

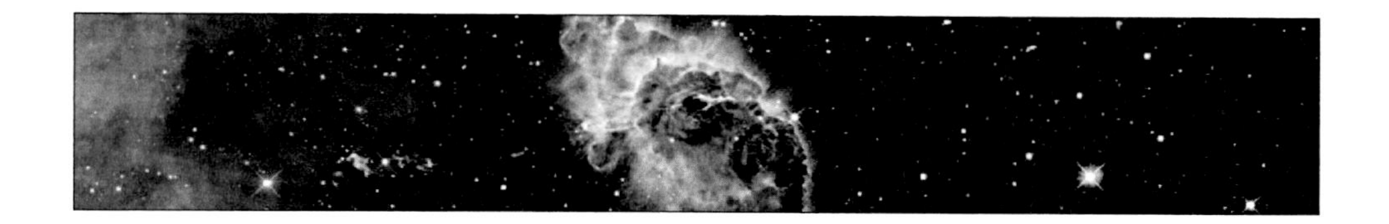

Segment 5

Does the Bible Conflict with Evolution?

Objectives

Students will learn:

1. Why the Bible does not conflict with the scientific theory of evolution,
2. Why a literalistic approach to reading the Bible causes confusion and leads to errors in interpretation and understanding,
3. The four main lines of evidence for evolution,
4. The significance of evidence from near-death experiences in supporting belief in the afterlife, and
5. The possibility of alien life — and perhaps intelligent alien life.

Overview

Joe is back with Alana and Dan to report on Tyler's newfound openness to the creator. However, Tyler is still stuck on one issue—evolution. In fact, he says, "If you want me to believe, answer the question about evolution." Alana and Dan explain that since the Bible doesn't conflict with science, it can't conflict with evolution either. The problem, they explain, occurs when people read the Bible literally, as a science book, and then don't want to believe anything other than the details of the stories, as if they would be disobeying God. They share some of the evidence for evolution with Joe, admitting that there is still much that is not explained, like how inanimate things became living things, and how living things became humans, with souls. They go on to share some evidence from near-death experiences that demonstrates there's something more than our material bodies which makes us think and feel, meaning our souls cannot be the product of evolution alone and even intelligent alien life, if it exists, could be endowed with souls by God.

Scientific Summary of Segment 5
Does the Bible Conflict with Evolution?

1. What is evolution?

Evolution is a scientific theory that explains the development of species using evidence from fossils, genetic similarities between species, and geographic distribution of species.

While it may not contain a complete explanation of the development of different species, among most scientists it is the accepted naturalistic explanation for the development of higher order species.

2. Does the Bible conflict with evolution?

Only if naturalistic evolution (which does not allow human beings to have a soul capable of surviving bodily death) is believed to be a complete explanation of the origin and development of all the species on earth, including humans.

At first glance, evolutionary theory seems to be in conflict with the Biblical account of creation where God makes man as a separate entity in his own image apart from the animals. Can the two accounts be reconciled? Yes. As we saw in the previous segment, the Bible is not exploring science, but rather theology, and the theological point in the Genesis account is that human beings are distinct from other animals and are made in the image of God. Thus, the Catholic Church has indicated that evolutionary theory is not necessarily in conflict with its theology.

Catholics may believe—or not believe—in evolution to whatever degree they wish (based on the best scientific information available, of course) up to and including the development of a physical organic brain, so long as they do not exclude the existence of a soul in human beings or claim that the soul is just a product of evolution.

This means that evolution cannot be a complete explanation for human beings because evolutionary theory is only concerned with biological processes, and the human soul is not biological, and, therefore, could not have evolved. Because it is immaterial, the soul also survives bodily death.

3. Is there any scientific evidence for the human soul?

Yes, there is much evidence for this in a testable phenomenon called near-death experience (NDE) which shows that it is likely that human consciousness survives bodily death. The studies of near-death experience are reported in top peer-reviewed medical journals and in independent peer-reviewed studies. They report on thousands of cases, with four factors scientifically corroborated in several careful case studies giving objective evidence for the survival of human consciousness after bodily death:

a. **Patients' reports of what is going on in the room where their body is lying (and even in rooms outside their location) can be verified after the fact because they have reported unusual data which would not ordinarily be part of a resuscitation procedure, such as:**

   (1) Cracks in light fixtures,
   (2) Conversations among people in another room,
   (3) Placement of dentures and shoes, and
   (4) The clothing of strangers inside and outside of the operating room.

   These accounts occurred during the period when the patient was clinically dead, with a flat EEG; with no electrical activity in the cortex; and with loss of brain stem function evidenced by fixed, dilated pupils and absence of the gag reflex.

   In a major study, Janice Holden surveyed 107 cases from 39 different publications by 37 different authors or author teams. She found that only 8 percent involved some inaccuracy.

   In contrast, 37 percent of the cases (almost five times as many) were determined to be accurate by independent objective sources, such as the investigation of researchers reporting the cases. The other 55 percent of cases did not have inaccuracies but could not be independently verified according to Holden's stringent criteria, but these cases are very difficult to explain unless human consciousness survived clinical bodily death.

b. **Most blind people report being able to see after clinical death.** Studies have shown that up to 80% of blind people who had a near death experience were able to see when they were clinically dead – many of them saw for the first time.

c. **Studies of near death experiences of children found overwhelmingly that they did not experience any fear of death after a near death experience, while other children, who had not experienced a near death experience, continued to have a normal or high death anxiety.**

d. **A 1982 Gallup poll discovered that about eight million people in the United Stated had had a near death experience.** They reported the following ten characteristics of their experiences:

(1)	Life review	32%
(2)	Being in another world	32%
(3)	Feelings of peace, painlessness	32%
(4)	Out of body	26%
(5)	Accurate visual perception	23%
(6)	Encountering other beings	23%
(7)	Audible sounds or voices	17%
(8)	Light phenomena	14%
(9)	Tunnel experience	9%
(10)	Precognition	6%

In view of these four factors and the circumstances under which they have been tested, there appears to be reasonable and responsible evidence for the survival of human consciousness after death. This would suggest a soul in human beings which cannot be explained by evolution. Even if the human brain is completely the product of evolution, there is a dimension of our self-consciousness capable of sight, hearing, thought, and feelings after clinical death which is not explicable in terms of the biological processes of evolution. Human beings seem to have both a body and a soul which are in union with one another prior to death; however, the soul exists separate from the body after death.

4. What about the existence of alien life elsewhere in the universe?

Given that there are 10^{22} stars in 10^{11} galaxies in our anthropic universe (designed for life), it is likely that there will be a large number of planets capable of sustaining life forms. This means that we cannot exclude the possibility of life forms on other planets. Though scientists are divided about alien life, it seems possible that other highly complex life forms may exist in our universe. If there were another life form on another planet which was capable of:

a. **Self-consciousness;**

b. **Self-transcendence;**

c. **An awareness of:**

(1) the infinite and eternal, and
(2) perfect truth, love, goodness, beauty, and being; and

d. **If such a life form had a similar capacity to that of human beings to survive bodily death;**

Then we would have to suppose that they would also have a soul which did not evolve within the physical universe alone, but came from God. If such beings existed we would be expected, as Christians, to evangelize and baptize them if they were not aware of love as the meaning of life; God being unconditional love; and Emmanuel (God with us) having come into our midst in a perfect act of unconditional love.

Handout 5a—Video Review and Discussion

Does the Bible Conflict with Evolution?

Opening Prayer

Father, we know that many Christians are confused by the seeming differences between the Bible's creation stories and science's theory of evolution. Help us to see the truth that you have revealed to us, through the use of our intelligence in scientific endeavors, and through faith in your word, so that we may be assured of your constant presence in the world. Amen.

Opening Reflection and Sharing

What do you think: *Does the Bible conflict with evolution?*

Review Questions

1. What issue is keeping Tyler from believing in the Bible?

2. What causes many Christians not to believe in evolution?

3. What are the four different kinds of evidence for evolution offered in the video?

4. What isn't the theory of evolution able to explain?

5. Why can't evolution account for our souls?

Name: **Date:** **Period:**

6. What was the most striking finding of *The Lancet*'s near-death experience studies?

7. What evidence is there that near-death experiences are real?

8. What one thing do persons having near-death experiences have in common?

9. What two groups of people give even stronger evidence for the conclusions about near death experiences?

10. If there is intelligent alien life elsewhere in the universe, how would it be possible for them to have souls?

11. What conclusions does the group reach about the evidence for near-death experiences?

12. What does the group say their next project will be to help youth strengthen their faith?

Class Discussion Questions

1. What caused Tyler to say that he would believe in the Bible if someone would answer the questions about evolution?

2. What do you think causes some Christians to want to believe that every word in the Bible is literally true?

3. Which evidence of near-death experience is most convincing to you? Why?

4. How has your view of the relationship between science and faith in God changed as a result of watching these videos and reflecting on them?

Closing Prayer

Loving God, we are grateful that you give us so many reasons to believe in you. You are at work in the world, constantly creating through the marvels of nature. You are at work in our lives, touching us through others and through our own experiences. May we always be willing to see you at work and believe in your presence in our lives. Amen.

Handout 5b

The Catholic Church and Evolution

Part One

Directions: Read the following quotes and answer the questions below.

The Church does not forbid the theory of evolution concerning the origin of the human body as coming from pre-existent and living matter – for Catholic faith obliges us to hold that the human soul is immediately created by God – be investigated and discussed by experts as far as the present state of human sciences and sacred theology allows.

Pope Pius XII, Humani Generis, 1950.

1. What distinction is the Pope making about what Catholics should believe and should not believe about evolution?

2. For what reason should choosing to believe in the evolution of the body be made?

Today, almost half a century after the publication of the encyclical (Humani Generis), new knowledge has led to the recognition of the theory of evolution as more than a hypothesis...(However), theories of evolution which, in accordance with the philosophies inspiring them, consider the soul as emerging from the forces of living matter or as a mere epiphenomenon of this matter, are incompatible with the truth about man. Nor are they able to ground the dignity of the person.

Pope John Paul II
Address to the Pontifical Academy of Sciences, 1996.

3. What does the Pope say new scientific knowledge has led to?

Name: **Date:** **Period:**

4. Which kind of theories of evolution are incompatible with the truth about human beings?

We are not some casual and meaningless product of evolution. Each of us is the result of a thought of God. Each of us is willed, each of us is loved, each of us is necessary.
Pope Benedict XVI, Inaugural Homily as Pope, 2005

5. Is the Pope saying anything different from his predecessors here, or is he repeating the same ideas? Explain.

It is not the case that in the expanding universe, at a late stage, in some tiny corner of the cosmos, there evolved randomly some species of living being capable of reasoning and of trying to find rationality within creation, or to bring rationality into it. If man were merely a random product of evolution in some place on the margins of the universe, then his life would make no sense or might even be a chance of nature. But no, Reason is there at the beginning: creative, divine Reason.
Pope Benedict XVI, Easter Homily, 2011

6. What is the Pope saying did not happen?

7. What would it mean for us if we were merely a random product of evolution?

Part Two

Directions: Give a response to each of the following statements from the perspective represented by the the recent popes quoted above.

I do not believe in evolution because it goes against what the Bible says about God creating each kind of living thing directly.

Response:

Because the mind is clearly a part of the brain, I think that consciousness, including our spirits or souls, evolved naturally as our brain capacity increased.

Response:

Name: **Date:** **Period:**

The Reason: Segment 5 Quiz

Does the Bible Conflict with Evolution?

Modified True or False

If the answer is true, mark true, but if it is false, mark false and re-write the sentence to be a true statement.

Example:

______ A. ***The Reason*** is a video series that attempts to demonstrate that science and faith are incompatible.

Answer:

False A. ***The Reason*** is a video series that attempts to demonstrate that science and faith are ~~incompatible~~. compatible.

______ 1. The Bible does not conflict with evolution.

______ 2. Some Christians reject evolution because they think it conflicts with what the Bible says about God's creation.

______ 3. There is no actual evidence for evolution.

______ 4. Evolution has been able to explain everything about the origin and development of all life on earth.

______ 5. Our souls, which are spiritual, can't be explained by evolution, which is only a physical process.

______ 6. Researchers have concluded that near-death experiences are the result of physical or medical causes.

______ 7. Several persons with near-death experiences, including the blind, have reported seeing or knowing things that should have been impossible for them to perceive. These claims have been verified independently.

_____ 8. Those who have had a near-death experience have a stronger fear of death.

_____ 9. If there is intelligent life on other planets, it's impossible that they may have souls.

_____ 10. Because of the evidence of near-death experiences, among other things, it is reasonable and responsible to believe that God exists, that our souls will live beyond our body's death, and that we have a divine and eternal purpose.

Meet the Scientists

Pim van Lommel

Dr. Pim van Lommel, M.D., (b.1943 in The Netherlands) completed his specialization in cardiology in 1976. He has researched the phenomena of Near Death Experience (NDE) since 1986 and is the author of more than 20 articles and a book, and has been a contributor to many publications on the same topic. In 2001, he and his colleagues published a study on NDEs in the prestigious British journal, *The Lancet* concluding that: "The NDE is an authentic experience which cannot be attributed to imagination, psychosis or oxygen deprivation."

Melvin Morse

Dr. Melvin Morse, M.D., is an associate professor of pediatrics at the University of Washington, recognized as one of the top pediatricians in the country. He began researching NDEs in 1982 after reviving a young girl who had no heart beat for 18 minutes. She recounted many details of what took place during the time she was close to death. At first skeptical, Morse has come to believe these are real experiences of human consciousness, not dependent on bodily function.

Notes

The Reason: **Final Discussion and Assignment**

Answering Agnostics' Questions

Directions: Imagine you are having a conversation with Tyler, Joe's roommate from the videos. Answer each of his questions based on the content of the series in complete sentences.

Segment 1: Can Science Disprove God?

Tyler—*How can you believe in God since science has already proven that he doesn't exist?*

Response 1—By describing the scientific method and its limits:

Response 2—By explaining why it is practically impossible for science to disprove the existence of physical things and impossible to disprove the existence of non-physical things:

Response 3—By describing the evidence for the beginning of the universe and what that means about creation:

Segment 2: Is There Any Evidence for a Creator in the Universe?

Tyler—*Okay, even though I accept that science can't disprove God, there's no way science could ever give us any reason(s) to believe in him. Science is the only real way we can know something is true, so anything philosophy tells us can be disproven by science. What do you have to say about that?*

Response 1—By describing three philosophical proofs for God's existence:

Name: **Date:** **Period:**

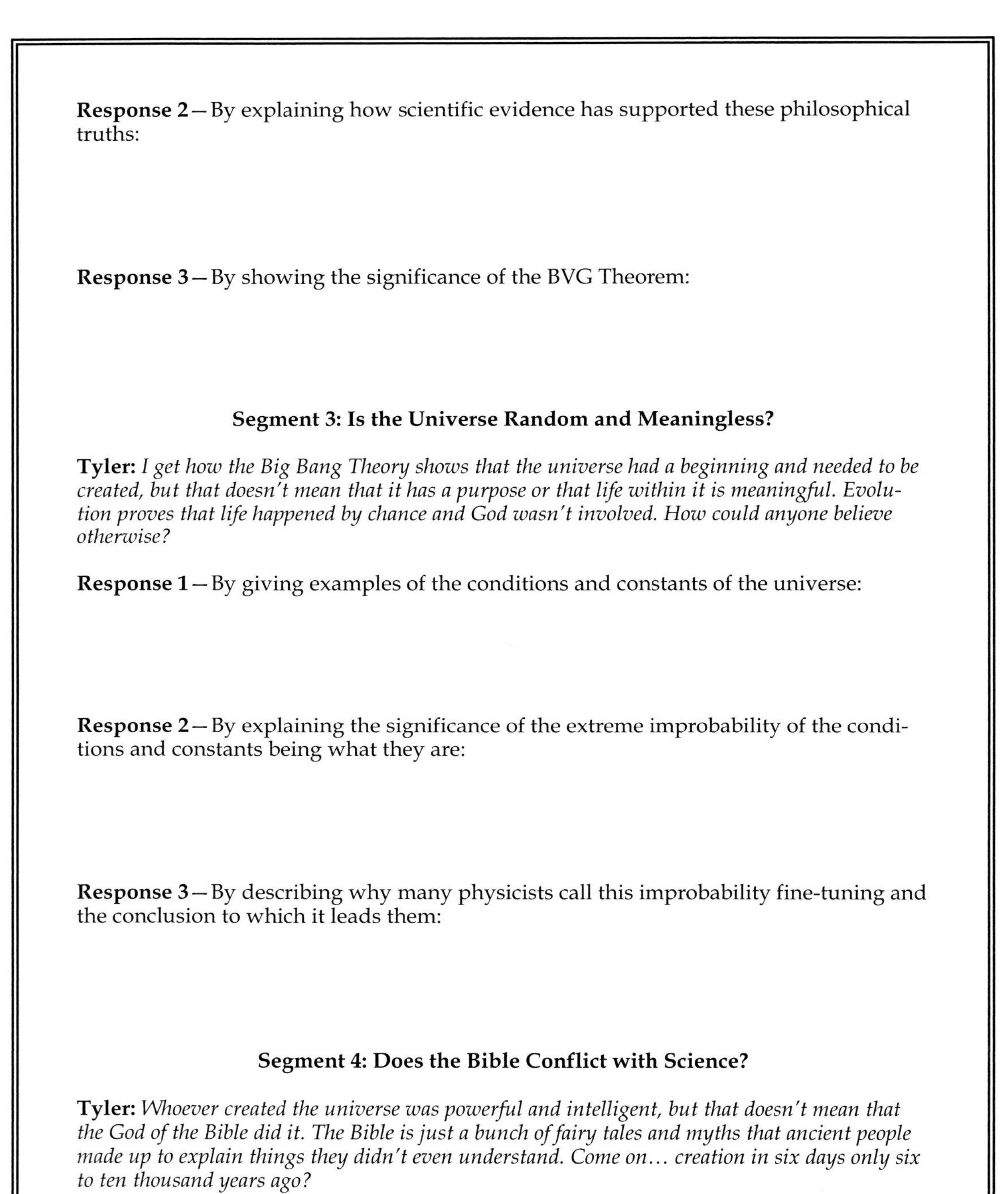

Response 2 — By explaining how scientific evidence has supported these philosophical truths:

Response 3 — By showing the significance of the BVG Theorem:

Segment 3: Is the Universe Random and Meaningless?

Tyler: *I get how the Big Bang Theory shows that the universe had a beginning and needed to be created, but that doesn't mean that it has a purpose or that life within it is meaningful. Evolution proves that life happened by chance and God wasn't involved. How could anyone believe otherwise?*

Response 1 — By giving examples of the conditions and constants of the universe:

Response 2 — By explaining the significance of the extreme improbability of the conditions and constants being what they are:

Response 3 — By describing why many physicists call this improbability fine-tuning and the conclusion to which it leads them:

Segment 4: Does the Bible Conflict with Science?

Tyler: *Whoever created the universe was powerful and intelligent, but that doesn't mean that the God of the Bible did it. The Bible is just a bunch of fairy tales and myths that ancient people made up to explain things they didn't even understand. Come on… creation in six days only six to ten thousand years ago?*

Response 1 — By describing the kind of book the Bible is and isn't:

Response 2 — By explaining the limits of what science and philosophy can tell us about the creator.

Response 3 — By showing why it makes sense that God would have revealed more about himself to us than we could know from science and philosophy:

Segment 5: Does the Bible Conflict with Evolution?

Tyler: *Look, I'll believe in the Bible if you can answer this one question: evolution. How come science gives us so much evidence for something that the Bible doesn't even talk about?*

Response 1 — By explaining why the Bible can't conflict with evolution:

Response 2 — By demonstrating that a literal reading of the Bible, when that was not intended by the authors, can give us an interpretation that conflicts with science:

Response 3 — By showing the significance of near-death experience in our understanding of the Bible and evolution.

Name: **Date:** **Period:**

Final Reflection: Share something you have learned from watching ***The Reason*** segments and participating in activities and class discussions that wasn't addressed in this final assessment.

COLLEGE LEVEL LECTURES AND COURSES

Are you interested in deepening the rationale for your faith from science and philosophy – or do you aspire to help others be more secure in their faith? Ideal for college students, high school seniors or homeschooling students seeking a contemporary intellectual defense of faith (for college credit or noncredit).

Check out these three online college level lecture series. These courses are designed for students without a background in science and philosophy, explaining basic logic and important scientific concepts in an accessible way. They use an online blackboard which is like being in the front row of a classroom. A preview is available on the Magis website.

1. **The Evidence for God from Physics and Philosophy** ($30.00) by Fr. Robert J. Spitzer, S.J., Ph.D. This non-credit lecture series explains Fr. Spitzer's award winning work – *New Proofs for the Existence of God: Contributions of Contemporary Physics and Philosophy* (Eerdmans 2010). It consists of 32 online lectures (approximately 35-40 minutes each). It can also be taken for two academic credits at Benedictine College as RI 298a for $215.00.
2. **The Evidence for Jesus Christ and Theological Anthropology** ($30.00) by Fr. Robert J. Spitzer, S.J., Ph.D. This non-credit lecture series explains Fr. Spitzer's upcoming book – *A Philosopher Looks at the Evidence for Jesus*. It consists of 32 online lectures (approximately 35-40 minutes each). It can also be taken for two academic credits at Benedictine College as RI 298b for $215.00.
3. **Physics and Metaphysics in Dialogue** ($45.00) by Fr. Robert J. Spitzer, S.J., Ph.D. This non-credit lecture series is the equivalent of an upper division undergraduate course in Philosophy of God (Philosophical Theology). It explains Fr. Spitzer's book *New Proofs for the Existence of God: Contributions of Contemporary Physics and Philosophy* as well as the problem of suffering, the evidence for a human soul and the problem of atheism. Important: this course significantly overlaps RI 298a above. It can also be taken for three academic credits at Benedictine College as GS 298 for $325.00.

MAGIS CENTER
OF REASON AND FAITH

College Credit:
When students enroll in the above lecture series for credit, they will be given reading assignments from the above books, exercises, study questions, study exams and real exams. They will also be provided with an email through which to ask questions.

For more information or to register, visit WWW.MAGISREASONFAITH.ORG and click "College Course"

MAGIS CENTER OF REASON AND FAITH

"Magis" (pronounced mah-jis) in Latin means "more." In some contexts it can mean "the even more" – the furthest frontier – frontiers of reason, knowledge, the universe and reality itself.

OUR MISSION
To explore and share the close connection between reason and faith as revealed by new discoveries in physics and philosophy.

OUR GOAL
To explain the consistency between faith, physics and philosophy.

CONTACT
Magis Center of Reason and Faith
2532 Dupont Drive
Irvine, California
92612 USA

Tel. (949) 271-2727
Fax. (949) 474-7732
Email. jjacoby@magis.us

Fr. Robert J. Spitzer, S.J., Ph.D., the presenter for In The Beginning: Evidence For God From Physics, *is the Co-Founder and President of the Magis Center of Reason and Faith.*

For more information, visit WWW.MAGISREASONFAITH.ORG

A Catholic Follow-up To The Alpha Course

Catholicism 201

www.catholicism201.com

Embraced for its ability to catechize and entertain, *Catholicism 201* has been run in more than 23 countries worldwide.

Although Fr. James Mallon designed *Catholicism 201* to directly follow the final talk of *The Alpha Course* entitled, "What About the Church?," this eight-week course is also being used as a stand alone program or as a component of sacramental preparation and RCIA.

The DVD course features eight talks by Fr. James on topics at the heart of the Catholic faith: The Church, Sacraments and Sacraments of Initiation, The Eucharist, Sacraments of Healing, Sacraments of Vocation, Mary and the Saints, Introduction to Christian Morality and The Thorny Issues. Each talk is divided into two parts approximately 30 minutes in length which enables *Catholicism 201* to be run as either an eight or 16 week course.

Complemento católico para el Curso Alpha

Catolicismo, 2ª parte **天主教信仰 201**

www.jp2mi.ca

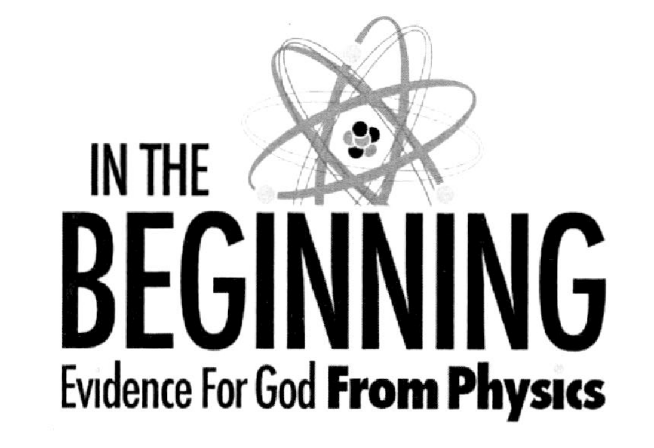

IN THE BEGINNING

Evidence For God **From Physics**

www.inthebeginningcourse.com

There is no contradiction between faith and reason. This encouraging truth is the bedrock for *In the Beginning: Evidence For God From Physics.* Ideal for the inquiring mind, this two-disc DVD set features four presentations by Fr. Robert J. Spitzer, S.J., Ph.D. which range in length from 48 minutes to just over an hour.

A former president of Gonzaga University from 1998-2009, Fr. Spitzer has authored several books including *New Proofs for the Existence of God: Contributions of Contemporary Physics and Philosophy*. The *In the Beginning* course explores the contents of this book including what, if any, the limits of scientific evidence are and whether such evidence provides support for a beginning and design of our universe.

The Big Bang Theory is addressed as is oscillating universe theory and the multiverse theory. Fr. Spitzer also speaks to the following questions: is the Bible doing science and can human beings be explained by an evolutionary (biophysical) process alone? The final week of the course explores evidence for survival of human consciousness after death as well as the possibility for extraterrestrial life.

www.magisreasonfaith.org